NOTES ON HORSE MANAGEMENT IN THE FIELD.

PREPARED IN THE VETERINARY DEPARTMENT

FOR

GENERAL STAFF, WAR OFFICE.

1919

(This supersedes WO $\frac{40}{2466}$ Catechism of

'tc.)

www.firesteppublishing.com

FireStep Publishing
Gemini House
136-140 Old Shoreham Road
Brighton
BN3 7BD

www.firesteppublishing.com

First published by the General Staff, War Office 1919.
First published in this format by FireStep Editions,
an imprint of FireStep Publishing, in association with
the National Army Museum, 2013.

www.nam.ac.uk

ISBN 978-1-908487-77-3

Cover design FireStep Publishing
Typeset by FireStep Publishing
Printed and bound in Great Britain

Please note: *In producing in facsimile from original historical documents, any imperfections may be reproduced and the quality may be lower than modern typesetting or cartographic standards.*

NOTES ON HORSE MANAGEMENT IN THE FIELD.

PREPARED IN THE VETERINARY DEPARTMENT

FOR

GENERAL STAFF, WAR OFFICE.

(This supersedes 40/WO/2466 Catechism of Animal Management, etc.)

INDEX.

(21767) 7365/P.P.2522. 15M. 1/19. M. & S. G.S.38.

PAGE

S.S. 646.]

NOTES ON HORSE MANAGEMENT IN THE FIELD.

I.—GENERAL.

1. **Introduction.**—These notes are based on experience in France and on the official publication, "Animal Management," which it is recommended should be studied.

2. **Resourcefulness Necessary when Difficulties Arise.**—The adoption of counsels of perfection is often impossible on Service. Resourcefulness must be exercised to meet difficulties.

Common sense and ingenuity will help to overcome troubles, and forethought to prevent their development.

3. **Need for System, Discipline, Training and Supervision.**—The O.C. a unit is responsible for the condition of the horses in his charge. He must arrange for the proper training of young officers, N.C.O.s and men in stable duties. Watering, feeding and grooming of horses must be carried out on a system understood by all ranks, and discipline must be such that duties are as thoroughly carried out when officers are absent as when they are present.

Large units must be sub-divided for purposes of supervision in stable duties into smaller units according to the arm of the Service. These smaller units should not exceed 40 horses.

4. **Objects aimed at by Horse-mastership.**—Horse-mastership aims at keeping the largest number of animals as fit as possible, and reducing inefficiency to a minimum by prevention of accidents and illness.

II.—WATERING.

1. **Quality and Quantity.**—Every effort should be made to ensure a clean supply of water in sufficient quantity.

Eight to ten gallons a day per animal is required.

2. **Need for Troughs.**—Water from troughs, if possible, and if regular troughs are not forthcoming, some substitute should be found. Anything capable of holding water will do, *e.g.*, biscuit and tea tins, ground sheets and tarpaulins of all kinds, variously supported above the level of the ground, if possible.

3. **Animals not to be led into Water.**—Unless absolutely necessary, animals should not be taken into the water to drink, particularly into ponds and muddy bottomed sluggish streams.

4. **Danger of Dirty Water.**—Drinking dirty water, if persisted in, will upset digestion and lead to sand colic and general inefficiency. It is also a common source of intestinal worms.

5. **Troughs—How Filled.**—Pumps should be used, if available. Take care not to stir up mud and sand while pumping. Lower the inlet pipe quietly into the water in a bucket, and keep it there while pumping.

6. **Water should be as near Camp as Possible.**—Time and energy spent in going to and from water is pure waste. The nearer the camp to its water supply the better.

Horses should not have to cross a main traffic road to get to their watering place.

7. **Watering in Camp.**—In standing camps water at least three times daily, and in summer four times.

8. **Watering on the March.**—While on the move allow a good drink whenever possible, however hot and sweaty the horses may be, unless very severe or fast work is expected. On the march mounted officers, N.C.O.s or men must be sent forward to reconnoitre watering possibilities at the next halt, so that watering may be undertaken the moment the unit halts, in the best manner circumstances allow.

9. **Not Immediately after Food.**—A heavy drink soon after a meal is liable to upset digestion, and should not be allowed.

The usual routine is water first, feed afterwards. At liberty horses drink as they feel inclined.

10. **Rules for Watering at Troughs.**—Proper water discipline at troughs is one of the most important items in stable management, and should always be supervised by an officer or N.C.O. One man should not lead more than two horses. He must stand in between them, and on no account must horses' heads be tied together. Horses should be formed up in column of sufficient width to allow one yard per horse at a trough. They should be dressed up by word of command, and no horse allowed to leave trough until the last horse has finished drinking. When horses have finished, the command "Right (or left) turn, file away," should be given, and the next batch of horses dressed to within a few yards of the trough, and on whistle, or command "Dress up," all horses are moved up to the trough. Every batch requires from five to six minutes at the trough.

Bits should always be removed and girths slackened.

Water troughs should be frequently emptied and cleaned.

Kickers should be watered separately.

11. **System of Decentralization Advisable.**—Drinking water ffers a ready means of spreading disease. It is better to provide eparate troughs for each unit than to allow one trough to be sed by a number of units.

Where several units have to water at one place, arrange a ime table to prevent overcrowding and kicking.

Any horse with the slightest suspicion of cold or infection of ny kind should be watered in its own canvas bucket, which hould be kept marked and separate. These buckets should not be illed by dipping them into the water troughs.

III.—FEEDING.

1. It is essential that animals should get the food to which hey are entitled.

2. There are certain recognized principles in feeding which nust be observed:—

(*a*) Water before feeding.

(*b*) Feed in small quantities three times, and preferably four imes, per 24 hours with grain and chaff, and twice with hay.

(*c*) If expecting hard work immediately after feeding, only ive a half-feed.

(*d*) Put a double handful of chaff in every feed.

(*e*) Keep nose bags or other feeding utensils clean.

(*f*) Use hay nets.

(*g*) Separate hay well out of bales, and damp it with salt and water, when available, before putting in hay nets. This should e done on tarpaulin or in a clean place.

(*h*) Mash horses once a week when possible.

(*i*) Put salt in feeds (horse ration 1 oz.).

(*j*) Graze whenever possible and encourage men to cut grass nd bring it into lines for horses. This may be mixed with the ay ration, adding to the bulk of food.

(*k*) Have nose bags taken off as soon as horses have finished eeding.

(*l*) Watch your horses feeding and know which are the slow eeders.

(*m*) Never give the order "Nose bags on for 5 (or 10) minutes." The proper order is give "Quarter (half or full) feed," nd each man must know the time required by his own horse to nish its feed.

(*n*) Never allow grooming kit or anything else, except horses' feeds, to be carried in nose bags.

(*o*) Men in charge of wagons must always take feeds and water buckets on the wagon when going on a journey, and must report on arrival whether they have watered and fed.

(*p*) Never feed in nose bags if you can improvise mangers or boxes.

(*q*) Feed and hay up as late as possible at night and give horses small feed of hay at or before Reveille.

(*r*) Chaff should be cut as short as possible.

(*s*) Save hay seed, short hay and clover leaves.

(*t*) When practicable, crush oats. *Crushed oats are more easily digested **than whole oats.***

3. In the absence of chaff cutters, one man can soon produce a useful amount of chaff with a heavy knife or chopper, using any log of wood as a chopping block. Rough grass from banks, hedgerows or fallow land is useful for the purpose.

A good improvised chaff cutter can be made by fixing a sickle horizontally in a biscuit box. The handle of the sickle should be on the outside of the box, the back of the blade is pushed down through a slit cut vertically in the end of the box, cutting edge or blade upwards. Additional rigidity is given by fixing the handle outside the box with wire. The man then sits or kneels on the back of the box, takes the hay fairly tight in both hands and pulls it towards him, thus chaffing the portion in one hand. By this method one man can cut sufficient chaff for 30 horses in three-quarters of an hour.

The difficulty of carrying chaff cutters can be got over, in the case of the lighter types of cutter, by carrying them under the driver's seat of a G.S. wagon. The feed block should be to the front and the driving wheel immediately in rear of the driver's box. The stand must be taken off and carried separately. The cutter should be bolted on to a spar lashed across the wagon. On the march this method has the advantage of providing a clean place (the inside of the wagon) into which to cut chaff the moment the wagon is unloaded.

4. **Scale of Ration.**—This is laid down in General Routine Orders, Q.M.G.'s Branch, and unless the issue is restricted by the military situation the full amount should be drawn.

Under war conditions it is frequently necessary to make use of whatever is obtainable, and the following may with advantage be given:—

(*a*) As a substitute for hay.—Oat, wheat, barley or pea straw, in the form of chaff if possible.

(*b*) As a substitute for oats.—Maize, small quantities of barley, linseed cake, linseed, peas and beans.

(*c*) As a laxative diet and to make bulk.—Bran, turnips, beetroot, mangolds, carrots, green crops, brewers' and distillers' grains.

5. **Wastage: Nose Bags, Hay Nets.**—Waste of food must be prevented at all costs.

Feeding off the ground should not be allowed. Both hay and oats at all times may become soiled, and in wet weather trodden into the ground. In windy weather hay is blown about and wasted.

Every animal should have a nose bag and hay net.

Nose bags, after use, should be turned inside out, cleaned and dried in the sun if possible. They should never be left lying on the ground.

6. **Save Hay Seed, Short Hay and Clover Leaves.**—Hay bales should be broken up and hay nets filled at a central place.

Economy will be effected if the hay is rubbed over an improvised sieve. Seeds, clover leaves and short hay will fall through and only the clean long hay should be put into the nets. The material which falls through the sieve, and would ordinarily be wasted, should be given as chaff.

The sieve can be made with hay or rabbit wire, secured to a rough wooden frame 5 or 6 feet long by 2 or 2½ feet wide, supported horizontally 2½ feet from the ground.

A clean paulin, sheet, blanket or strip of latrine canvas should be placed on the ground wherever forage is stored or chaffed or feeds made up. This will prevent the waste of oats, hay seeds, &c., that would otherwise be trodden into the ground.

7. **Do not Feed Mouldy Forage.**—Mouldy hay or corn does more harm than good, and should not be given.

Bran mashes are useful, but bran is not absolutely necessary for horses at work. It is chiefly used for horses which are sick or those at rest.

For a bran mash take 2 to 3 lbs. of bran, a tablespoonful of salt, and as much *boiling* water as will well wet the bran. Cover and allow to stand till cool enough to be eaten.

8. **Epsom Salts and other Drugs.**—Epsom Salts and other drugs are unnecessary, and may be harmful. Horses no more require regular dosing than do men.

9. **On the March.**—On the march the Staff arrange certain halts. O.sC. horsed units must decide, on receipt of march orders, where they will water and feed, and what proportion of feeds will be consumed at each halt.

It is better to give a quarter or half feed at several halts than a whole feed at one halt. In any case only under urgent military necessity should horses be allowed to go more than four hours without food.

10. **Effects of Wrong Procedure.**—The over-hungry horse is so impatient that he tosses food about and wastes it, and will bolt his food without masticating it properly. Wind-sucking, dung-eating, and other objectionable habits are largely due to leaving animals too long without food or to giving an insufficiency of bulk.

11. **Grazing and Green Food.**—Grazing must be carried out systematically. The grazing guard, consisting of 1 N.C.O. and 4 men as a minimum, should first be posted, 1 man on each side of the ground it is proposed to graze. The horses (or mules) should then be led to the centre of the grazing ground and knee haltered with a head rope or rein. The men, other than the grazing guard, should then walk quietly away. The men of the grazing guard should walk slowly up and down their side of the ground and keep the animals from straying. The N.C.O. i/c will be responsible that the orders for grazing are properly carried out. He will see that all horses are properly knee haltered before the men who have brought them out go away. If, during grazing, knee halters slip down below the knee he will readjust them.

All men in mounted and transport units should be taught to knee halter as follows:—

> Fasten one end of the head rope to the lower dee of the backstrap of the head collar, pass the other end round one of the horse's legs above the knee, make a clove hitch, shorten the standing part of the rope, making a stop to prevent the hitch tightening, finishing off the spare end with half hitches on the upper dee of the backstrap.

Knee haltering is the best method of securing horses at graze; they are soon accustomed to it and are also able to lie down and roll, which is most beneficial.

In addition to regular grazing for several hours under a grazing guard as above, men in charge of horses should graze their horses on road sides and banks at every opportunity.

The grazing on banks is improved by burning the old coarse 'rass in the early spring. Horses prefer short sweet grass to long oarse grass. The latter should be cut with a clasp knife and rought into the lines in hay nets and mixed with long hay to ncrease bulk.

In the devastated area, where lucerne, clover and sainfoin may e available, these should not be grazed, but cut and brought into he lines and fed green in hay nets. Lucerne, especially, can be ut repeatedly, whereas, if grazed, the growth ceases.

The practice of tying animals together to graze must be orbidden.

IV.—GROOMING.

1. **Cleanliness the First Object.**—It is the duty of officers nd N.C.O.s to train their men as thoroughly in grooming as in nusketry or any other part of their profession. Good grooming revents casualties from skin troubles. Once a man has been roperly taught, he grooms quickly and well, whilst a badly rained man tires himself and does little to improve his horse.

Grooming must be carried out in a workman-like manner, vith jacket off, sleeves rolled up, spurs removed, and braces own (every mounted soldier must have a belt). Quick, hard rooming, is what is required. A man must put his will and veight into it. No horse is groomed properly which is not roomed quickly. A good man should groom and clean thoroughly dirty horse within an hour. Horses should be groomed in the un whenever possible.

The primary object of grooming is to clean the skin and to revent disease, but the general health is much improved thereby.

Do not suppose that animals on active service, and in the pen, require no grooming.

Grooming must be systematic.

Give special attention to head, throat between jaws, tail, and nder chest. These are often neglected and become starting oints of skin diseases.

Do not wash legs. When wet, dry them to prevent cracked eels.

Wash sheaths occasionally.

Place rug or blanket on a sweating horse if he cannot be ried quickly.

2. **Examination of Feet.**—Horses' feet should be picked out vhenever the horse returns from exercise or work, and at the eginning of each stable hour.

3. **Grooming Tail.**—A good tail greatly adds to the appearance of a horse. When cleaning it the hair should be parted and brushed out from the roots lock by lock for cleanliness, and finally straight down for appearance. On no account should men use a comb on the tail.

Horses' tails should not be cut and mules' tails should not be clipped.

4. **Stable Routine (Grooming at Mid-day Stables).**—

(*a*) Pick out feet.

(*b*) Examine horse and report any casualty to your N.C.O., who will at once report it to the officer.

(*c*) Wash out eyes, nostrils and dock with moistened rubber.

(*d*) Groom with body brush the neck (starting behind the ear), shoulders and foreleg in succession, then the body, belly, hindquarters and leg.

Begin on the near side, and then take the off side in the same order.

When grooming on the near side always use your brush in the left hand, and when on the off side use brush in the right hand.

(*e*) Turn horse round. Groom head, neck and chest. Tie up. Brush out tail. Hand rub or wisp for ten minutes. Clean head collar. Then report to N.C.O., "Horse ready for inspection."

5. **Spare Horses.**—In cases where there are a great many spare horses and it is not possible to groom them all thoroughly at mid-day stables, some may have to be left over to be groomed at evening stables; these will be groomed at mid-day stables next day, and so on, thus ensuring that all horses get thoroughly groomed at least once every other day. It is better to groom horses well every other day than to groom them badly every day.

6. **Time for Grooming.**—Length of time is no criterion of quality of work done, and quick hard grooming should be encouraged in every way. Men should be allowed to clean harness as soon as they have finished grooming.

Horses must on no account be groomed while feeding, and idle men who have not cleaned their horses by the time the order to feed is given should be made to do extra grooming in the afternoon.

7. **Grooming Tools.**—The grooming tools supplied on Service are:—

> Dandy brush, when specially demanded.
> Body brush.
> Hoof pick (on clasp knife).

These must be kept in serviceable condition, and any deficiency at once reported.

Wisps are made by the men as required, and every man should be taught how to do this.

8. **Use of Body Brush.**—The body brush is for removing scurf and dirt from horse's coat.

To use brush with the best result, the man should stand well away, keep his arm stiff and lean the weight of his body on the brush.

If the man stands close to his horse, with bent elbow, and brushes with his arm only, he does not force the bristles or fibres of the brush through the coat so well or remove the scurf so effectually.

In grooming the belly, face the horse's tail and apply the brush the way of the hair with the back of the hand down, elbow up, brush in left hand on near side and right hand on off side.

9. **Use of Curry Comb (to be improvised locally).**—The curry comb is for cleansing the body brush of scurf, and is useful for removing caked dirt from coat. When grooming on near side, curry comb should be in right hand, and *vice versa*, the hand passed through the webbing at back of comb.

An occasional rub on the brush is all that is required, say once in every 3 or 4 strokes of the brush.

Lazy men make a great show of cleaning the brush, and do little work on the skin, the result being that the brush is quickly worn out.

Dirt should not be dislodged by blowing scurf out of comb.

The contents of comb should be knocked out into squares made on the ground behind horses. This has the advantage of showing at a glance the amount of work the man has accomplished. This scurf should be swept up on turning out from stables.

10. **Use of Wisp.**—Wisping is really a form of massage. It is a most valuable method of improving the condition of the skin and coat and for making muscle. It stimulates the skin generally and improves circulation.

The wisp should be brought down with a bang on the skin in the direction of the hair, and the process repeated all over the body.

The only disadvantage of the wisp is that it has to be made out of a horse's hay or straw ration.

11. **How to Make a Wisp.**—A wisp is a pad of hay or straw made by twisting the material into a rope and doubling it into convenient sized pads as described below.

To make a wisp, soft hay or straw should be twisted into conveniently sized rope about eight to ten feet long. Two loops are then formed at one end, one being very slightly longer than the other. Each of these loops in turn is then twisted beneath the remainder of the rope until the end is reached, when it is passed through the extremity of each loop and tucked under one of the twists.

A really good wisp should be no wider than can be conveniently grasped by the hand; about a foot long and two or three inches thick. Some little practice is necessary to make one really well, and great care must be taken to see that hay or straw is not wasted in making it. Material for the purpose is given by the N.C.O. The wisp should be damped before using. (*See* diagram.)

A Wisp.

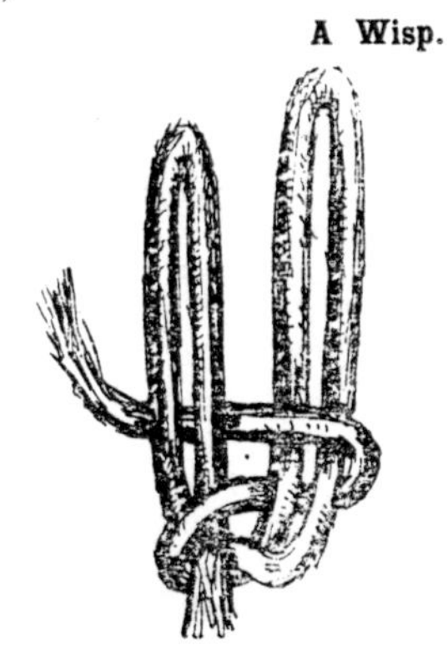

FIG. (a).

Commencement, showing two loops formed from one end of rope, and method of twisting the other around them.

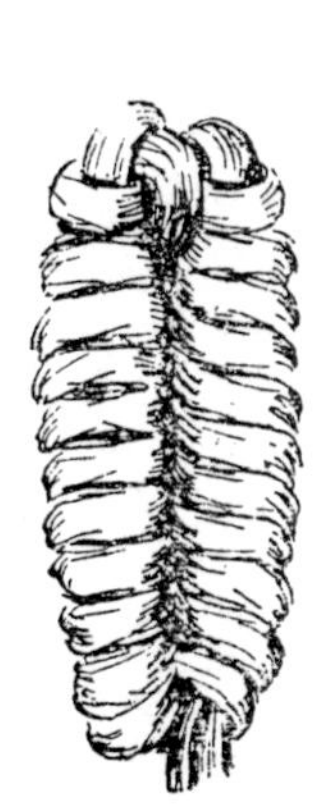

FIG. (b)

Completed.

12. **Hand Rubbing.**—Is an excellent form of massage. It is specially useful for removing the loose hair of the coat, as well as stimulating the skin, and has the advantage over the wisp that it does not reduce the horse's forage.

The hands are slapped down smartly on the coat one after the other, and the weight of the body leant upon them and the forearms, whilst both are moved over the skin with firm pressure. It originated in the East, and is of the greatest value in keeping horses in condition. For removal of loose hairs by this method, hands and arms should be slightly damp.

Hand rubbing the legs and "stripping" the ears, *i.e.*, pulling them gently through the hand, from base to apex, are both details which should not be neglected, and in the case of the legs brisk but gentle rubbing with fingers and palms in the same direction as the hair is an excellent stimulant to the circulation of the limbs.

13. **Detection of Skin Disease, &c.**—Officers and N.C.O.s during stables should instruct men and at the same time keep a close watch on the animals.

The men should be trained to notice while grooming anything unusual either in the behaviour of the animal or in the appearance of the skin, and to report signs of itchiness, lice, nits, ringworm spots, patches where hair has come out, and any other irregularities.

On Service, skin diseases are liable to break out. Early detection is most essential. Failure to detect the trouble until it has spread extensively in the unit indicates inexcusable negligence.

14. **Precaution against Disease.**—It is of the utmost importance that grooming kit, rugs and nosebags should be properly marked and on no account changed; this changing is one of the surest ways of spreading disease.

15. **Sand Baths.**—Sand baths have many advantages, and unless mange or other skin diseases exist they should be encouraged. It is good for an animal to roll, and mules particularly appreciate it.

V.—PICKET LINES.

1. **Best Method.**—Picketing may be either on breast-high air rope, secured between wagon wheels, trees or posts, or on a ground rope secured by means of picketing pegs. The former is the better method.

2. **Rope to be Secure and Taut.**—The rope in both cases should be strong, well secured and kept taut.

3. **Single Peg Objectionable.**—Single picketing pegs are objectionable, unless the head ropes are kept short, and heel ropes are also used. They are necessary for picketing kickers and for single horses generally, when it is best to shackle the horse's fore fetlock and secure it to the peg by a rope 1-ft. to 18-in. in length. This method allows the horse to turn his quarters to the wind, avoids the possibility of rope galls, and makes it very difficult for the horse to pull the peg out of the ground.

4. **Need for Short Head Ropes.**—Never allow the head rope to be too long. Heel galls are caused by getting the hind leg caught in a too long head rope.

5. **Prevention of Heel Galls.**—For animals on a ground line, the head rope should be just long enough to allow the head to be held in a natural position over the line. On a breast line the head rope should be of sufficient length to admit of the animal getting its head to the ground.

6. **Advantages of Heel Ropes.**—Heel ropes, whatever may be said against them, reduce the chance of serious injuries from kicking.

The head and heel ropes should be just taut when the animal is standing in a natural position square to the line.

Kickers are best dealt with by hock hobbles. A hobble or shackle should be fastened just above each hock and connected together by a rope, strong strap or chain, about 8 inches long.

7. **Position of Picketing Lines.**—Place picketing lines at right angles to, rather than parallel with, contours. This ensures a more or less level standing for each horse whatever the slope may be.

8. **Drainage of Lines.**—Attend to the drainage of lines, particularly in relation to storm water, by improving existing drains, or making new ones, as may be necessary. Do not wait for bad weather.

Forethought here may frequently save trouble later.

9. **Stable Guards.**—Post stable guards for duty between stable hours and at night, with instructions for the care of animals placed under their charge.

Their duties are:—

(*a*) To keep the head and heel ropes tied up at the proper length.

(*b*) To shorten up the nose bags of any horses that are tossing them up.

(*c*) To remove nose bags as soon as the horses have finished feeding, turn them inside out, and collect them.

(*d*) To note and report horses off their feed.

(*e*) To keep hay nets and hay in proper position for horses to eat with comfort, and to prevent the latter from being trampled under foot.

(*f*) To remove droppings, and keep the lines clean.

(*g*) To prevent the horses becoming frightened, by speaking to them, and at the least signs of a stampede to call for assistance.

10. **Standings.**—The objects, in order of importance, are to keep horses out of the mud, the wind, and the rain. This must be borne constantly in mind when making plans for new, or improvement of old, standings.

Wind screens can be formed out of hurdles made of brushwood, gorse, reeds, straw, &c. When available, rabbit wire is useful as a support for wind screens.

Anti-bomb traverses should be 6 feet high, 3 feet thick at top, and 7½ feet thick at bottom. They then act as efficient wind screens in addition to giving protection against bombs.

Where rivetting material is not available they can be made either (*a*) with plain earth covered with any rough turf procurable, in the form of an Irish bank, or (*b*) with alternate layers of earth and dung well pressed down and finally covered with road scrapings, clay if available, or earth. If the fresh dung is properly covered with earth sods flies will not breed in it.

Anti-bomb traverses should be erected so as to leave a passage 3 feet wide behind the horses, and proper provision should be made for drainage of storm water.

The plan should be spitlocked on the ground before commencing work and battens or posts and string used to keep the work in progress level.

Standings should be as close to metalled roads as possible. Their length should run at right angles to the contours, so as to give each horse a more or less level standing and assist drainage. Drains should be cut before the horses are placed in the standings. The floors can then be improved with any hard material available. If unbroken bricks can be procured, they should be regularly laid and fixed between battens or any suitable timber placed at intervals.

Cinders on top of chalk are very satisfactory.

Standings should always be raised well above the level of the ground, and slightly sloped for drainage. If no hard material is available the earth excavated from the drains should be rammed on top of the existing surface, before the standings are taken into use.

Roofs.—The best roofs are made with corrugated iron, but the material must depend on what is available. Rabbit wire and tarred felt or paper may be used, but should never be used for walls as they are too easily broken. Care should be taken that the roofs project far enough to protect the horses' quarters.

The site for stabling should be near a road and good approach roads to standings should be made. Firm approach roads to water troughs are also necessary.

VI.—HEALTH, CONDITION, EXERCISE, &c.

1. **Appearance of Health.**—The following are indications of health:—Head alert, eyes bright, ears pricked, appetite good, body well furnished, skin supple and bright, standing even or resting one hind leg, droppings fairly firm and not slimy, urine light yellow and rather thick in appearance.

2. **Condition.**—Condition is not merely a question of looking well. It means ability to do some special work satisfactorily with a minimum of strain.

It must not be confounded with fatness. It can only be secured by a gradual process of preparation, and attention to detail.

Each class of animal should be, as far as possible, brought on gradually at its own work, particularly draught horses and remounts recently received.

When work is insufficient, exercise for from 2½ to 3 hours a day; heavy draught horses at a walk and others part walk and part trot.

An Officer should attend exercise parties, and a N.C.O. should be detailed to front and rear of each party to regulate pace and to prevent straggling.

Care should be taken to bring back animals cool to their lines.

An animal in soft condition sweats easily and very soon becomes tired and distressed. If work under these circumstances be forced, complete exhaustion and even death will follow.

VII.—SHOEING.

1. **Supervision of Shoeing.**—The Officer Commanding a unit hould carefully watch his animals' feet. Much inefficiency arises rom faulty shoeing and preventible foot injuries and ailments.

Daily inspections should be made by a farrier or shoeing mith.

The rate of wear of shoes varies, but they should be replaced s worn out.

No animal should be allowed to go over one month without ttention to feet and re-shoeing, the feet being lowered, and ither a new set of shoes put on or the old shoes re-applied termed a " remove ").

" Bent up toes " are recommended for horses which stumble r wear out shoes very quickly.

2. **Mules' Feet often neglected.**—It is very common to find he feet of mules much neglected and allowed to grow too long.

3. **Cold Shoeing.**—Though " cold shoeing " is the recognised nethod on active service, every opportunity should be taken for hoeing hot.

4. **Picked-up Nail.**—Animals working on ground long in ccupation of troops are liable to very serious injury from " picked-p nails." Travelling kitchens and camp fires where wooden boxes re burnt cause free scattering of nails. Every supply or refilling lump or building being constructed adds to the danger.

Control in this respect is most essential.

All should be encouraged to pick up nails and to put them n specially provided receptacles.

VIII.—HARNESS AND SADDLERY.

1. **Harness Fitting.**—The fitting of harness requires constant ttention. The principal points to be looked at are the position f the breast collar, the adjustment of the pole chains, breeching, nd neck and loin supporting straps.

The position of the breast collar will depend on the neck upporting strap, the latter should neither be too short or too ong; if too short the strain in draught comes mainly on the top f the neck, if too long the draught falls below the point of the houlder, in either case galling usually results.

The pole chains require careful adjustment; if too short, the animal does not get squarely into the collar; if too long, the pole is not acted on when the animal hangs back in the breeching. Similar remarks apply to the adjustment of the breeching; if too short, the freedom of movement is restricted, while if too long, the pole is not acted on.

The loin straps supporting the trace should not be too short, otherwise they will take some of the strain of draught, and will gall the loins.

Harness should be kept clean, soft and pliable.

2. **Saddle Fitting.**—The special points to be attended to are:—

(*a*) No weight can be borne on the top or sides of the withers or along the central line of the back.

(*b*) The shoulder blades should be allowed freedom of movement.

(*c*) Weight should be evenly distributed over the upper surface of the ribs extending from the play of the shoulder backwards.

The backs of riding and pack horses should be inspected after every march. Every gall has its particular cause, which must be traced, and where possible remedied without delay.

Instruction in folding the saddle blanket is necessary, as the intelligent employment of this article of equipment often permits animals which have begun to lose flesh being safely worked.

The importance of avoiding sitting in the saddle for long periods at a time should be impressed on all ranks. Dismounting at every halt and occasional spells of dismounting and walking the men on a long march benefits men and horses.

3. **Clothing.**—At least one rug per animal is necessary in the open during winter. If rugs cannot be carried, one must depend on the saddle blanket.

A rug during rain, even if wet through, is an advantage.

It is advisable that rugs should be numbered, to ensure the same one being constantly used for the same animal.

IX.—CLIPPING.

Unless the troop horse can be clipped before the end of November, there is a risk of his feeling the loss of the protection afforded by the coat during the winter season.

If left unclipped, there is danger of widespread mange and ertainty of extensive lousiness.

To obtain the best results, it is advisable to clip the whole ody before the end of November, and subsequently to clip trace igh only.

The legs should not be touched in any circumstances.

When severe lice infection has taken place, clipping may be ecessary as part of the treatment.

The only practical way of dealing with clipping is by means f power clippers. Instructions for the care of these can be btained from Ordnance Service.

Appendix A.

INSPECTION OF SHOEING.

Attention should be directed to the following points:—

1. **The Preparation of the Foot.**—Before the shoe is fitted the foot is to be properly lowered. This should be done with the rasp; the knife should not be used on the frog or sole except to remove any ragged or exfoliating horn. The outside of the wall should not be rasped.

2. **The Fitting of the Shoe.**—The shoe should be fitted flush all round the edge of the hoof except at the heel, where a little extra width is permitted, especially in heavy draught horses. The shoe should extend to the heel, but should not project appreciably beyond it. It should have an even bearing on the wall, bars and outer margin of the sole.

3. **Nailing on and Finishing.**—The nails should be driven regularly and evenly at a sufficient height to secure a good hold, those at the toe being a little higher than those at the quarters. The clip and clenches should not project beyond the crust.

Appendix B.

EXAMPLES OF FEEDING.

	Heavy Draught.		Other Horses over 15 hands.		Horses under 15 hands.	
	Oats. lbs.	Hay. lbs.	Oats lbs.	Hay. lbs.	Oats. lbs.	Hay. lbs.
Reveille	—	2	—	1½	—	1
Morning	3	2 (C)	2	1½ (C)	1½	1½ (C)
Mid-day	4	2 (C)	2½	1½ (C)	2	1½ (C)
Afternoon ..	—	1½	—	1½	—	1½
5 p.m.	4	2 (C)	2½	1½ (C)	2	1¼ (C)
8 p.m.	4	2 (C)	3	1½ (C)	2½	1¼ (C)
Hay up	—	2½	—	2	—	2½
TOTALS ...	15	14	10	11	8	10

NOTE.—(C) = Chop.

NOTES.

NOTES.

NOTES.

NOTES.

NOTES.

NOTES.

NOTES.

NOTES.

NOTES.

NOTES.

NOTES.

NOTES.